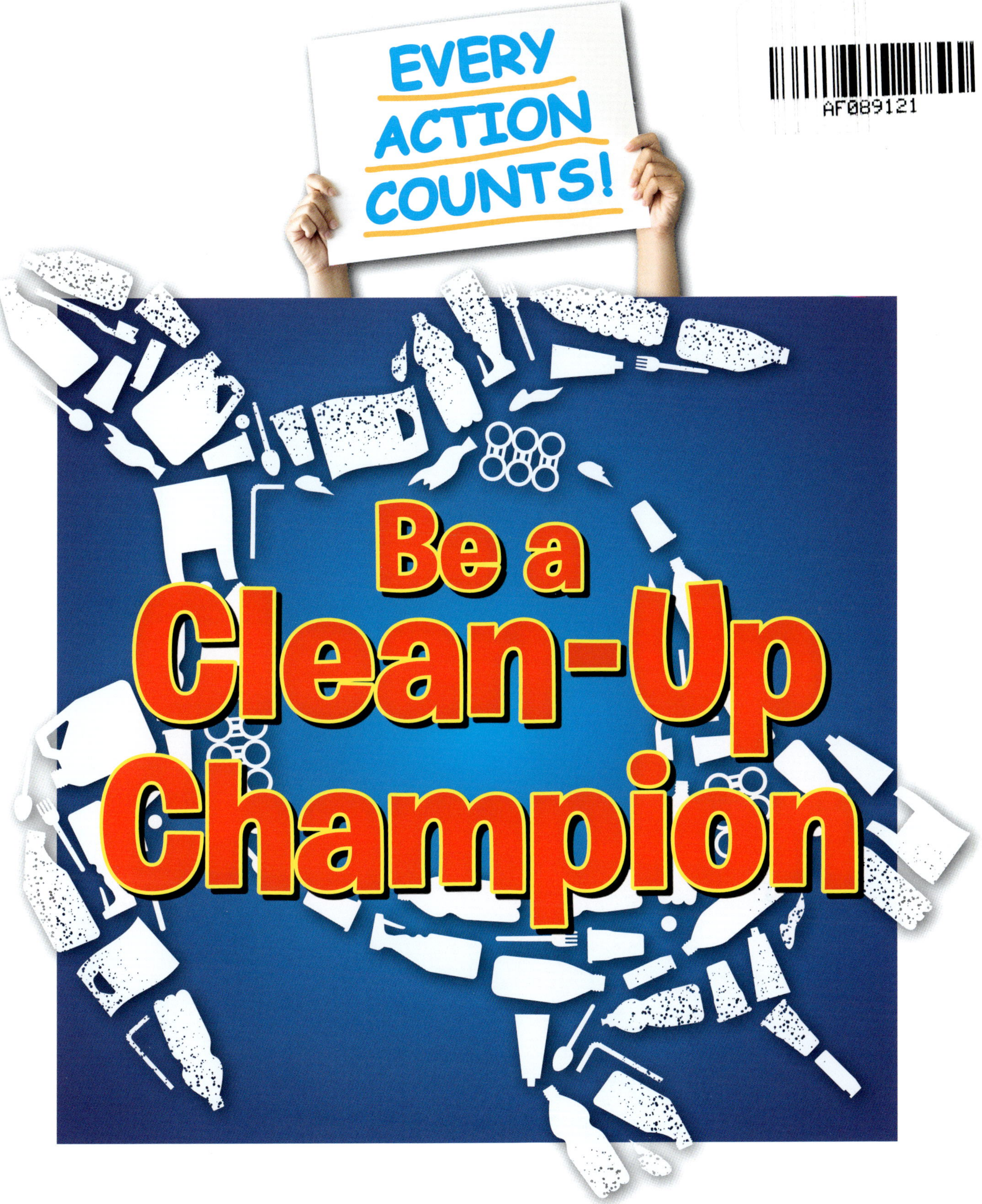

Be a Clean-Up Champion

By Belinda Gallagher

Ruby Tuesday Books

Published in 2024 by Ruby Tuesday Books Ltd.

Copyright © 2024 Ruby Tuesday Books Ltd.

All rights reserved. No part of this publication may be reproduced in whole or in part, stored in any retrieval system, or transmitted in any form or by any means, electronic, mechanical, photocopying, recording, or otherwise, without written permission from the publisher.

Editors: Ruth Owen & Mark J. Sachner
Production: John Lingham

Photo credits:
Alamy: 31L (Matt Matthews); Ruby Tuesday Books: 9B, 11T, 15T, 23L; Shutterstock: CoverTL (Kristine Rad), CoverTR (KlingSup), CoverBL (Breslavtsev Oleg), CoverBR (Motortion Films), 4T (Fedelya), 4CL (mmilan), 4CR (Malgosia Janicka), 4BL (Africa Studio), 4BR (Frans Blok), 5T (KlingSup), 5BL (Alfa Photostudio), 6T (David Pereiras), 6CL (Alan Morris), 6CR (BirgiCt Ryningen), 6B (Africa Studio), 7C (EvaL Miko), 7B (Wave Break Media), 8T (Aleksey Kurguzov), 8CL (amirraizat), 8CR (CGN089), 8B (Addictive Stock), 9TL (Marija Stepanovic), 9TR (Sasha Chornyi), 10T (Kristine Rad), 10BL (oasisamuel), 10BR (Hayran 1), 11BL (Alexisaj), 11BR (Tom Gowanlock), 12T (Rich Carey), 12CL (Breslavtsev Oleg), 12CR (xalien), 12B (Visual Art Studio), 13 various, 14T (Dalibor Danilovic), 14C (photka), 14B (Little Kid Moment), 15BL (Elena Chevalier), 15BR (Akhmad Dody Firmansyah), 16T (Alohaflaminggo), 16CL (Lucia Zanmonti), 16 various, 17 various, 18T (Damron Rattanapong), 18C (Zamrznuti tonovi), 18B (Studio Romantic), 19TL (Norenko Andrey), 19TR (New Africa), 20T (Nikola Fific), 20C (Monkey Business Images), 20B (James Jiao), 21T (Pressmaster), 21C (EQRoy), 21B (Pressmaster), 22T (Tatevosian Yana), 22C (mitsuap/Aksenova Natalya), 22B (Ami Parikh), 23TR (Caspian81), 23CR (sylv1rob1), 23BR (T.Dallas), 24TL (TSViPhoto), 24TR (Fevziie), 24C (Merrimon Crawford), 24B (AB-7272), 25 various, 26T (Monkey Business Images), 26CL (Prostock-Studio), 26 various, 27 various, 28TL (gob_cu), 28TR (chuchiko17), 28B (PeopleImages.com – Yuri A), 29TL (stokkete), 29TR (Motortion Films), 30T (Valentina Razumova), 30BL (Dragana Gordic), 30BR (Deemerwha Studio), 31R (Kletr).

ISBN 978-1-78856-442-7

Printed in Malta by Gutenberg Press Ltd.

www.rubytuesdaybooks.com

Note from the Publisher

Neither the publisher nor the author can accept legal responsibility or liability for any loss, harm or injury that may come about from following the instructions in this book. All activities should be carried out with adult guidance and supervision. Some activities involve being out of doors in public spaces. Children should be accompanied at all times. It is the parent's or carer's responsibility to ensure their child is safe.

CONTENTS

Are You Ready to Clean Up?........................4
Look Around You..6
Say No to Litter!...8
Clean Up for Wildlife.................................10
Keep Beaches Clean..................................12
Reuse and Recycle....................................14
Plastic Swaps and Drops...........................16
Swap and Repair.......................................18
Your Legs Can Keep Air Clean!..................20
Water Matters..22
Don't Waste Food.....................................24
Clean-Up Celebration................................26
You Are a Clean-Up Champion!.................28
Glossary...30
Index, Quiz Answers.................................32

Staying Safe!

All the activities in this book are fun and easy to do. Be sure to ask an adult to help you with each one at every stage. Never go anywhere without your trusted adult. Wear old clothes for the make-and-do activities. Always make sure an adult is nearby when using scissors or a glue gun. When outside, make sure you are wearing clothes that suit the weather. Have fun!

Are You Ready to Clean Up?

Our planet needs a clean up. Litter, plastic and **pollution** are making a massive mess.

Rubbish affects all of us, and it harms wildlife and natural places.

PLASTIC gets into rivers and the ocean and then washes up on beaches. Seabirds try to eat rubbish or get trapped in it.

LITTER ends up on streets and in parks.

CHEWING GUM sticks to the pavement. What a mess!

RUBBISH ruins natural habitats where animals live.

Did you know?

Single-use plastic is plastic that is used once and then thrown away, such as plastic cutlery. We should try our best not to use it.

However, if everyone took a few small actions, the problem wouldn't be so big.

We can all help to do something about litter and pollution.

Try taking one small action. Pick up one piece of litter and put it into a litter or recycling bin.

Small actions like this may not seem much. But imagine if your friends took the same action. That would soon add up to a lot of litter!

Make it Count

Next time you go to a park or other outside space, try doing a litter pick.

All you need is a reusable bag and some gloves.

Go to page 7 to see what you can do after your litter pick.

Look Around You

Have you noticed litter where you live? Can you spot litter when you walk to school or play outside? Sometimes we need to stop and look around us. Litter is everywhere.

Next time you are in a car, see if you can spot rubbish at the side of the road.

When you go for a walk in the park or the countryside, you might see plastic bags caught in trees and hedgerows.

In towns and on streets, you may see food or packaging dumped on the ground.

Litter Data Experiment

In a notebook, draw a chart to record information about the litter you pick. Write down your data, or findings, on the chart for one week.

What did you find?	Where did you find it?	What is it made from?	Can it be recycled?	How many did you find?
Bottle				
Bottle lid				
Soft drinks can				
Crisp packet				
Food wrapper				
Carrier bag				
Clothing				
Coffee cup				
Food				

When your experiment is over, look at your results. Here are some questions to think about. There are no right or wrong answers.

1. What kind of litter is dropped the most?
2. Where is most litter found?
3. Why do you think people drop litter?
4. How can people be encouraged to stop dropping litter?

There are many reasons why people drop litter. They may not be near a bin, or they don't want to carry it.

At picnic sites, some people may be too lazy to clear up their litter. It starts to pile up.

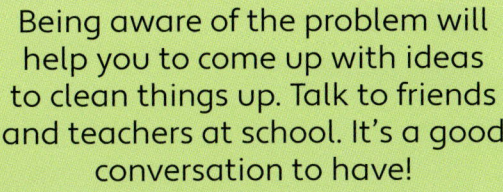

Being aware of the problem will help you to come up with ideas to clean things up. Talk to friends and teachers at school. It's a good conversation to have!

Say No to Litter!

The good news is there are lots of ways to tackle litter. Here are some simple actions you can take – encourage your friends and family to try them, too.

Never drop litter on the floor, put it in a bin. If you can't find a bin, take your litter home and put it in the bin.

When you're out and about, separate your litter into the correct recycling bins.

Plastic bags are bad for the planet. Say no to them! Always take a reusable bag with you on days out. They are great for putting litter in, too.

Did you know?

Plastic is a **material** made by people, it doesn't come from nature. When a plastic bag gets thrown away, it just breaks into smaller and smaller pieces and is never truly gone.

Try doing five-minute litter picks. You'll be surprised by how much litter you can pick in this time.

Top Tip

You can't pick up all the litter, that would be too hard. Remind yourself that every piece counts – you're doing a great job!

Clean up the Waves for Whales

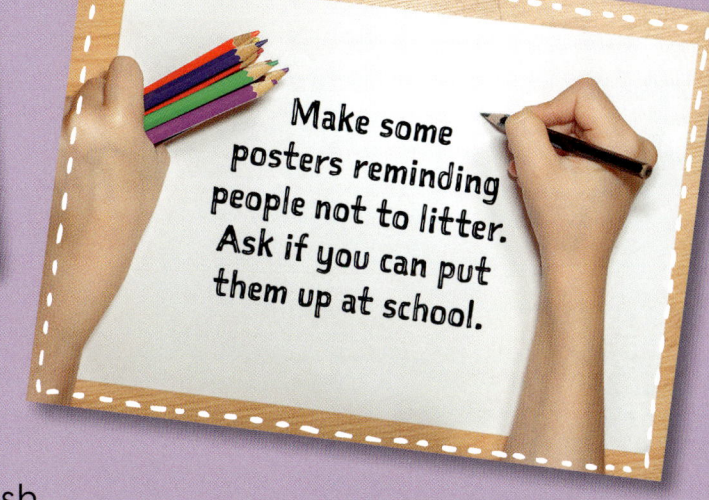

Make some posters reminding people not to litter. Ask if you can put them up at school.

Create a picture with some litter you've picked on a countryside walk or beach visit. Give your picture a title. Your litter art will remind people not to drop rubbish.

Litter-Picking Champions

Form some litter-picking teams at school or with your family and friends. Make your litter picks a competition with a winner's trophy made from recycled bottle tops.

You will need:
- Plastic cartons, lids and bottle caps
- A glue gun (and adult helper)
- Scissors
- 2 pipe cleaners
- Gold or silver spray paint
- Recycled newspaper

1. Design your trophy by stacking different-sized cartons and lids to make a trophy shape.

2. Ask an adult to help you glue the sections of the trophy together using a glue gun. Allow the glue to dry.

3. Take your trophy outside and place it on some newspaper. Spray paint one side of the trophy. Allow it to dry. Turn the trophy over and spray the other side. Allow it to dry.

You might need to add a second coat of paint.

4. Bend the pipe cleaners into handle shapes. Cut them to size if needed. Glue them to your trophy.

5. Decide a date and location for your first litter-picking competition. Remember gloves and bags.

Split into teams of two or three people. Decide how long the litter pick will last. On the word GO, start your litter pick. Who will be the litter-picking champions?

Clean Up for Wildlife

Plastic and litter are dangerous for wildlife. Birds may feed plastic to their chicks. Foxes may eat plastic by mistake. Bottles in ponds and lakes may trap frogs. Every piece of litter you pick up and recycle could save a wild animal.

Top Tip
Make sure the lids on rubbish bins are down, even the ones you see when you're out and about. This will stop wild animals from getting in and eating something that could harm them.

Even when you dispose of litter properly, it can still get into the **environment**. Rubbish might blow from a bin. A person who's emptying a bin might accidentally drop a piece of rubbish. Use these tips to make the litter you throw away as safe as possible.

1

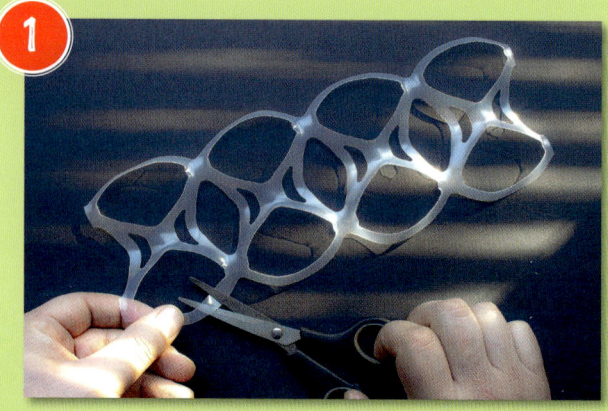

Help to prevent an animal from getting caught in plastic can holders. Cut them up with scissors before putting them in the bin. Do the same with elastic bands.

2

Ask an adult to squeeze cans shut. This will prevent small animals from getting inside.

Reusable T-shirt Bag

Make this reusable bag from an old T-shirt. It's great for litter picks. Wash and use again!

You will need:
- An old T-shirt
- Scissors

1. Lay the T-shirt out on a flat surface. Cut off the sleeves.

2. Cut the neck out of the T-shirt.

3. Now cut 5-cm-long slits at the bottom of the T-shirt. Make sure you cut through the front and back at the same time. You will end up with a fringe.

4. Tie a front and back piece of fringe tightly together in a double knot. Continue to tie all the fringe pieces together.

5. Your bag is ready to use! Make some for your friends and family, too.

3

Balloons are bad for wildlife. Animals may eat them by mistake. Never let a balloon go outdoors. After you have finished with it, pop the balloon, cut it up and put it in the bin.

4

Tie plastic bags in a knot before putting them in a bin. This will stop wild animals from getting trapped in them.

Keep Beaches Clean

Beach litter looks bad and sea animals may get trapped in it or eat it. People love to spend time at the beach, too. We must keep beaches clean so that we can all enjoy them.

Pick up litter and find a bin!

Did you know?

Litter often ends up in rivers. When rivers flow into the ocean, they carry litter with them out to sea. When the tide comes in, litter washes back up on beaches.

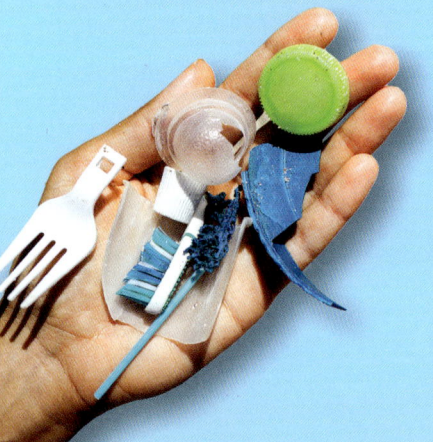

Use smaller pieces of litter on the beach to make some beach art. This shows how much we need to clean up. Take a photo if you can. Then litter pick your picture!

At the end of the day, check carefully that you have not left any litter behind.

Think about what you take to the beach. Take a packed lunch in a reusable container and a reusable drinks bottle. Then there will be no need to leave anything behind.

Be a Beach Scavenger!

Next time you visit the beach, have a scavenger hunt. See how many items you can find on this page or make up your own list.

- Seaweed — **5** points
- Seagull — **5** points
- 3 different shells — **10** points
- Worm cast — **10** points
- Footprints — **10** points
- Crab — **15** points
- Sea glass — **15** points
- Driftwood — **10** points
- Unusual-shaped pebble — **20** points

Give yourself points for each item you find.

Can you reach **100** points?

Collect any litter you see as you hunt, then put it in a bin afterwards. See how many items you can find on this list or make up your own.

Top Tip

Spending time in nature is good for you. Make sure your day at the beach is filled with fun and happiness.

Reuse and Recycle

Rubbish from our dustbins is taken to **landfill** sites to be buried. When we reuse and recycle, it means less of our rubbish ends up in landfill.

Landfill site

Did you know?

A landfill site is a huge hole dug in the ground. More rubbish is added until the hole is full. Millions of tonnes of rubbish are buried every year in the UK.

Many of the things we throw away can be recycled and made into new things.

Plastic bottles

Metal cans

Paper and cardboard

Glass bottles and jars

If you are not sure if something can be recycled, ask an adult to help you.

Did you know?

Glass bottles and jars can be recycled and made into new bottles and jars, over and over. Metal cans can be recycled again and again, too, and made into new cans.

Recycle old clothes, toys and games by taking them to charity shops. Then someone else can make use of them.

Make a Plastic Bottle Piggy Bank

This piggy bank is a fun way to recycle a plastic bottle.

You will need:
- A plastic bottle with a top
- 2 A4 sheets of craft paper
- Scissors
- A pencil
- A glue gun or double-sided tape
- An adult helper with a craft knife
- Acrylic paint
- A paintbrush
- A black marker pen
- Peel-and-stick googly eyes
- 4 matching bottle tops

Try making more for friends and family as gifts.

1. Cut a strip of craft paper that's long enough to go around the middle of the bottle. Draw a 3-cm-long rectangle in the centre of the strip.

2. Stick the paper around the bottle with glue or tape.

3. Ask an adult to cut out the rectangle to make a coin slot in the bottle and paper.

4. If you wish, paint the bottle top and allow to dry. Draw on the nostrils with the marker pen. Stick on the googly eyes.

5. Draw and cut out two ear shapes and a thin tail from the paper. Twist the tail around a pencil to make it curly.

6. Stick on the ears and tail. Finally, stick on four bottle tops as legs.

Add coins to your piggy bank. What will you save up for?

Cardboard tubes, egg cartons, plastic bottles and yoghurt pots are great for garden and craft projects.

Reuse glass jars and bottles as vases for flowers and for storing things like pencils.

Plastic Swaps and Drops

We use too much plastic and then throw it away. Plastic isn't a natural material. It takes thousands of years to break up into millions of tiny pieces called **microplastics** that will never disappear.

Did you know?

Micro means small. Microplastics are in our rivers, oceans, soil and even in the air. They get eaten by fish, birds and other wildlife.

There are lots of ways to swap or drop (stop using) plastic things. This will help reduce microplastics.

SWAP THE PLASTIC

Swap your plastic toothbrush for one made with **bamboo**. Bamboo is a plant, so it's a natural material that breaks down without harming the environment. You can even bury your old toothbrush in your garden!

DROP THE PLASTIC

Buy fruit and vegetables loose without plastic packaging.

Plastic wrap

Pop them in a reusable bag.

DROP THE PLASTIC

Special wraps made of beeswax can be used instead of plastic clingfilm to keep your lunch fresh. They can be washed and used again too.

Did you know?

Beeswax is a soft, waxy substance made by honeybees. They use it in their hives to make small spaces to keep their eggs and honey safe.

SWAP THE PLASTIC

Try using wooden pencils instead of plastic felt-tip pens for art and school projects.

Recycled glass jar

DROP THE PLASTIC

Use a soap bar instead of liquid soap in a plastic bottle.

Swap or Drop Challenge

With your family, each choose one plastic item you will try to swap or drop for one week. Keep a record of your challenge to see how well everyone does. Do you think you can add different items each week? Here are some ideas.

Kitchen sponge **SWAP** Washable cloth

Bread in plastic **SWAP** Paper-wrapped bread

Packaged sandwich **SWAP** Homemade lunch

Plastic coffee cup **SWAP** Reusable bamboo coffee cup

Crisps in a plastic packet **SWAP** Veggie snacks

Plastic honey bottle **SWAP** Glass honey jar

17

Swap and Repair

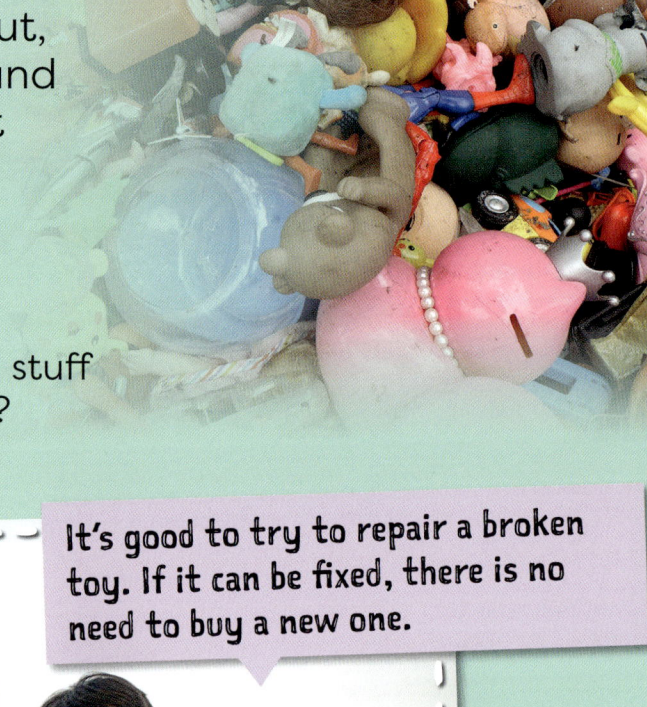

People buy too many things. If things get broken or worn out, they are often thrown away and replaced by new things. What happens to all the old stuff? It makes tonnes of rubbish that ends up in landfill sites.

How about if we swapped our old stuff and tried to fix the broken things? There would be a lot less rubbish.

It's good to try to repair a broken toy. If it can be fixed, there is no need to buy a new one.

It's fun to learn how to fix things.

Did you know?

Repair cafés and workshops are fun places to visit to see all kinds of things being fixed. You can learn new skills, reduce waste and save money!

Have a game or toy swap day with friends. This is a great way to try something new instead of buying it.

Can you and your friends start a book swap club? This is a great way to share a good story or cool facts.

Patch

Clothes can be repaired! Add colourful patches to jeans or shorts to repair holes. Or perhaps cut down a pair of jeans to make some summer shorts.

Make Some Patches

Ask an adult to help you cut old clothes up into patches. Clothes with pictures or patterns work well. Save these patches to add to shorts, jeans, leggings or pyjamas. Anything you like!

You will need:
- Old clothes with fun pictures or patterns
- Scissors

Keep your patches in a box with other scraps of material that might be useful or fun to use.

Your Legs Can Keep Air Clean!

It's true! If we walked, biked and scooted more, there would be less traffic and **fumes**. Cars, trucks and planes pump out dirty **gases**, or fumes. This causes air pollution, and it makes the air bad to breathe.

Go for a walk! Think about the places you visit. Can you walk to them instead of travelling by car? Write a list and talk to your family about walking instead.

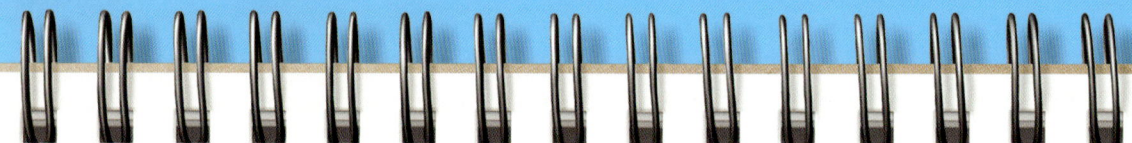

Start a Walking School Bus

Have you heard of a walking school bus? A group of schoolchildren and adult volunteers walk together as if they are on a bus. Children hop on board the walking bus from safe pick-up points.

Ask an adult if you can start a walking bus. Explain why it's a good idea. Here are a few reasons to talk about. Try thinking of some others.

- Better for the environment
- Good exercise
- Less traffic around the school
- Teaches road safety
- People spend time as a community

Use your bike or scooter. Going for bike rides is great fun and good exercise for our bodies. Plan a bike ride with your family for your next day out.

Top Tip

Can you share a ride? You need to travel in the car for longer journeys. But perhaps go with friends so only one car is needed. That means fewer vehicles on the road.

Did you know?

Buses and trains are cleaner ways to travel. They carry lots of people, which means less traffic.

Trees Help Clean Up the Air

To make their food, trees need a gas called carbon dioxide.

This gas is in the fumes pumped out by cars and other vehicles.

Trees suck the carbon dioxide and other harmful fumes out of the air with their leaves.

Then their leaves release clean oxygen, the gas we need to breathe!

Can you plant a tree?

Water Matters

Keeping water clean and safe for living things is an important part of being a clean-up champion.

Picking up litter will stop it getting into rivers and flowing out to sea. Think about the water you use at home, too. Don't waste it and be careful how you use it.

Think before you tip!

Only water should go down the plughole. Substances like cooking oil, paint and glitter can end up in rivers if you tip them down the sink.

Do you know adults who kill weeds in their gardens by watering them with poisonous **weedkillers**? Ask them to please stop! Weedkillers pollute the soil. Insects and other small animals may drink or be harmed by the poisonous water.

Wear gloves and offer to pull up the weeds instead!

Think Before You Flush!

Flush
- Pee
- Poo
- Toilet paper

Never Flush
- Plastic cotton buds
- Dental floss
- Wet wipes

All the things we flush flow along pipes to a **waste water treatment plant**.

Big sieve-like filters trap poo and other stuff. The water and pee flow through the filters.

The liquid is cleaned and made germ-free. Then it flows into a river or the sea.

A cotton bud may get through the filter. Then the cotton bud might flow out to sea, and an animal may eat it!

REMEMBER!
Only flush the 3 Ps

Pee   Poo   Paper

Save and Recycle Water for Plants

Use the boiled water from cooking pasta or vegetables to water plants. It has lots of goodness in it that plants love.

Let the water get cold before you use it.

Top Tip

Don't waste water. Turn off the tap when you brush your teeth. If you see a dripping tap, turn it off. Small actions like this add up.

Don't Waste Food

When food is thrown away, most of it ends up in landfill sites. This makes so much rubbish. How can we clean up our actions around food?

Did you know?
Food in landfill sites becomes rotten and gives off gases that pollute the air.

We should only buy the food we need. Plan a menu for the week ahead and make a shopping list. It's fun to look forward to your meals.

Don't scrape everything into the bin. Most leftover food can be frozen or kept in the fridge to eat the next day.

Rescue wonky carrots and potatoes at the supermarket! They might get left behind and thrown away. They taste just as good and shouldn't be wasted.

Food in tins and packets lasts for a long time. Ask to look in the kitchen cupboards. Are there foods that your family may never eat?

Did you know?

About one-third of all the food that is produced in the world each year gets wasted!

Donate the food to a **food bank** and stop it going to waste. The food will go to people who need it.

Fantastic Food-Saver Frittata

A frittata is a baked egg dish that can be eaten hot or cold. Ask an adult to help you make this yummy frittata using leftover vegetables.

You will need:
- Leftover vegetables – anything will work!
- 4 large eggs
- Handful of grated cheese
- A whisk and bowl for whisking
- A dinner-plate-sized ovenproof dish
- An adult helper with oven gloves
- A knife

1. Preheat the oven to 180°C.

2. Chop the vegetables into bite-sized pieces. Place in the ovenproof dish.

3. In a bowl, whisk the eggs and cheese together.

4. Pour the egg and cheese mixture over the vegetables.

5. Ask an adult to put the dish into the oven. Bake for 25 minutes until golden brown.

6. Check the frittata has set by inserting a knife. If not quite set, return to the oven for 5 minutes.

The frittata should be soft but solid. A bit like a baked cake.

Baked frittata

7. Allow the frittata to cool then cut into small portions.

Clean-Up Celebration

By now, you have discovered lots of ways you can help to clean up the world around you. Celebrate by inviting your friends to a clean-up party.

Plan Your Party

Here are some ideas for the perfect reusing and recycling party. Choose a date for your party, and give yourself plenty of time for planning.

1 Ask an adult to help email invitations to your friends. That way you won't need to use paper invitations.

To make bunting, cut out cardboard triangles. Fold over each triangle's short edge and tape it over string.

2 Make party decorations out of leftover craft materials, recycled cardboard and scraps of gift-wrapping paper.

Coloured, patterned paper can be used to make paper chains.

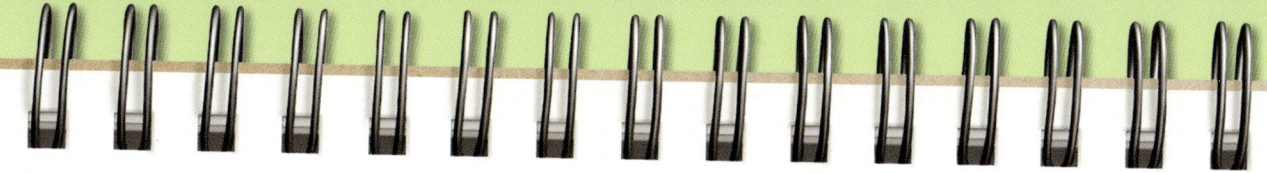

3 Chop leftover fruit and vegetables into little chunks and sticks. Serve them on plates or in containers that can be washed and used again.

4 Make the frittata on page 25. Cut it into small portions and serve cold.

Recycled jam jars

Serve with paper straws.

5 Make a delicious milkshake by whizzing up a cup of milk, an over-ripe banana and 3 ice cubes. Try other fruits, too.

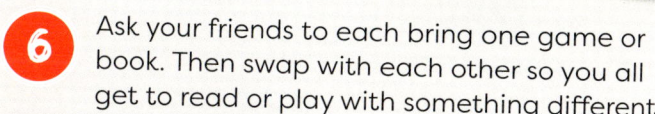

These bananas won't last long. Don't waste them!

6 Ask your friends to each bring one game or book. Then swap with each other so you all get to read or play with something different.

7 Share your clean-up ideas with your friends.

Top Tip

Talking about litter, reducing plastic use and recycling is just as important as cleaning up. Your voice is a brilliant clean-up tool.

You Are a Clean-Up Champion!

Every action you take to clean up is important, however small it may seem. By reading this book, you are now a Clean-Up Champion. Which of these clean-up actions have you tried?

PICK UP AND BIN LITTER
You are making the world a cleaner place!

REUSE OR REPAIR
Think before throwing things away. Could they be reused or repaired?

Top Tip
Pass this book on to a friend and ask them to do the same when they have read it. Let's create as many clean-up champions as possible!

CLEAN A BEACH
You are helping to keep oceans safe from pollution.

AVOID SINGLE-USE PLASTIC
Replace single-use plastic with reusable items.

RECYCLE

Sort rubbish and recycle all you can.

PLANT AND WALK

Help to clean up the air by planting a tree, and walking or biking.

Clean-Up Quiz

Take this fun quiz to see how much you remember from your book. The answers are inside the book and also on page 32.

1. What should you do if you can't find a bin to put your litter in?

2. What kind of bags should we use instead of plastic ones?

3. Should you do litter picks or litter drops?

4. Which of these things can't you recycle?

   Cardboard box   Plastic straw

   Metal can

5. Why is litter bad for wild animals?

6. Where is rubbish taken to get buried?

7. What might you do with your old clothes?

8. What should we plant to clean up the air?

9. What can we do instead of buying packaged fruit and vegetables?

10. How does walking help to keep the air clean?

There are many more champions, just like you. That means lots of people are helping to clean up our world – and that's amazing!

GLOSSARY

bamboo
A strong, woody, fast-growing grass. Bamboo is used for making many things, including shopping bags and reusable coffee mugs. It's natural, unlike plastic, and is easy to grow, so it's renewable.

environment
The natural world that surrounds all living things. We must try to keep the environment clean.

food bank
A place that people can visit to get free food. The food is donated, or given, by people who want to help others. Donated food also helps to reduce food waste.

fumes
A smoky mixture of harmful chemicals and gases. Traffic, factories and some machines create fumes.

gas
A substance that we cannot see or feel. Gases make up the air we breathe. Carbon dioxide and oxygen are types of gases.

landfill
A huge space where rubbish is buried in the ground.

material
What something is made of. Plastic, wood and paper are types of materials.

microplastics
Tiny pieces of plastic litter. Microplastics come from bigger pieces of plastic as they break up, especially in the ocean.

pollution
Harmful substances in the natural environment. Traffic fumes and litter are both types of pollution.

repair café
A place where people take things to be fixed. This is a good way to reuse our things instead of throwing them away.

waste water treatment plant
A place where the waste water from our toilets, sinks, baths, showers and washing machines goes to be cleaned. Solid stuff, such as poo, is removed and the water is cleaned. Then this water is released into rivers and seas.

weedkiller
A poisonous substance that is used to kill wild plants known as weeds. Weedkillers are often mixed with water so they can be poured or sprayed onto weeds.

INDEX

A
air pollution 16, 20-21, 24, 29
art 9, 12

B
bamboo 16
Be a Beach Scavenger! 13
beaches 4, 9, 12-13, 28

C
cardboard 14-15, 26, 29
charity shops 14
Clean-Up Quiz 29
clothing 7, 14, 19, 29

E
environment 10, 16, 20

F
Fantastic Food-Saver Frittata 25
food 6-7, 16, 24-25

G
glass 13, 14-15, 17

L
landfill sites 14, 18, 24
litter 4-5, 6-7, 8-9, 10-11, 12-13, 22, 27, 28-29
Litter Data Experiment 7
litter picks 5, 8-9, 11, 12-13, 29
Litter-Picking Champions 9

M
Make a Plastic Bottle Piggy Bank 15
Make it Count 5
Make Some Patches 19
metal cans 10, 14, 29
microplastics 16

O
oceans 4, 12, 16, 28

P
Plan Your Party 26-27
plastic 4, 10, 14-15, 16-17, 23, 27, 28
plastic bags 6, 8, 11, 29
plastic bottles 14-15
pollution 4-5, 20, 22-23, 28

R
recycling 5, 7, 8-9, 10-11, 14-15, 23, 26-27, 29
repair café 18
repairing 18-19, 28
Reusable T-Shirt Bag 11
reusing 11, 14-15, 19, 28-29
rivers 4, 12, 16, 22-23
rubbish 4, 6, 10-11, 14, 18, 24, 29

S
Save and Recycle Water for Plants 23
single-use plastic 5, 28
Start a Walking School Bus 20
Swap or Drop Challenge 17
swapping 16-17, 18-19, 27

T
traffic 20-21
trees 6, 21, 29

W
water 22-23
weedkiller 22
wildlife 4, 10-11, 12-13, 16, 22-23, 29

Page 29 Quiz Answers

1. Take your litter home with you **2.** Reusable bags **3.** Litter picks **4.** Plastic straw **5.** They might eat litter or get trapped in it **6.** A landfill site **7.** You can repair old clothes or recycle them at a charity shop **8.** Trees **9.** Buy loose fruit and vegetables **10.** Walking creates less traffic and fumes